TEACHING
MULTIPLICATION
USING
LEGO® BRICKS

Dr. Shirley Disseler

COMPASS

Teaching Multiplication Using LEGO® Bricks

Brigantine Media/Compass Publishing
211 North Avenue
St. Johnsbury, Vermont 05819
Phone: 802-751-8802
Fax: 802-751-8804
E-mail: neil@brigantinemedia.com
Website: www.compasspublishing.org

ORDERING INFORMATION

Quantity sales

Special discounts for schools are available for quantity purchases of physical books and digital downloads. For information, contact Brigantine Media at the address shown above or visit www.compasspublishing.org.

Individual sales

Brigantine Media/Compass Publishing publications are available through most booksellers. They can also be ordered directly from the publisher.
Phone: 802-751-8802 | Fax: 802-751-8804
www.compasspublishing.org

ISBN 978-1-9384065-5-3

CONTENTS

DEDICATION

To my sons Steven and Ryan, whose love of LEGO® bricks as children inspired me to find ways to use the bricks in education to engage young minds in math!

And to the 2016 graduate students in STEM at High Point University for being the test audience for my activities!

ACKNOWLEDGMENTS

Thanks to Neil Raphel and Janis Raye for their efforts in keeping me sane and moving forward throughout this project.

Much appreciation goes to Dr. Mariann Tillery, Dean of the School of Education at High Point University, for continuing support of all of my projects.

INTRODUCTION

Multiplication! Learning how to multiply can be daunting for young students. For most people, memorizing times tables is one of the least enjoyable tasks they have to do during math instruction in the elementary years.

And it's *not* the best way to learn multiplication. The rote process of repeating multiplication tables over and over, taking speed tests, and writing math facts ten times each—these instructional methods are not supported in educational research.

Students learn the concept of multiplication best through the process of modeling. Using manipulatives, they create models that show the relationship between the numbers. As they model the facts, their brains utilize creative and logical processes together. Research shows this to be the preferred learning format because it stimulates a "need to know" (Jensen 2005; Willis 2006; Persaud 2013).

This book offers a number of activities to help students understand the concept of multiplication. Some are useful when students are first introduced to multiplication, and others are directed toward learning how to multiply larger numbers. Typically, students learn the topics covered by the activities in this book in grades 2 – 5.

Students begin with the basics—understanding the meaning of multiplication as it relates to repeated addition. This understanding usually begins in grade 2 with the introduction of sets and its relationship to skip-counting. Following the

Jensen, Eric. 2005. *Teaching with the Brain in Mind, 2nd Edition*. Alexandria, VA: Association for Supervision and Curriculum Development.

Persaud, Ramona. 2013. "Education, the Brain, and the Common Core Standards." *Edutopia*. http://www.edutopia.org/blog/education-brain-common-core-ramona-persaud.

Willis, Judy. 2006. *Research-Based Strategies to Ignite Student Learning: Insights from a Neurologist and Classroom Teacher*. Alexandria, VA: Association for Supervision and Curriculum Development.

initial introduction of multiplication, students begin to work on basic facts of multiplication through modeling, arrays, and problem structure application, typically in grade 3. They work on one-digit multiplication problems and later matriculate to two-digit multiplication and beyond by grade 5.

These activities provide students with practice modeling the action of multiplication and support the engagement of students in understanding multiplication. Having fun with math helps enhance the desire to do math and encourages students' involvement in the learning process.

Why use LEGO® bricks to learn about multiplication?

LEGO® bricks help students learn mathematical concepts through modeling. If a student can model a math problem, and then be able to understand and explain the model, he or she will begin the computational process without struggling.

Modeling multiplication with LEGO® bricks is an easy way for students to demonstrate understanding of the vocabulary and the concepts of whole numbers. When students model the action of multiplication with LEGO® bricks, they have the opportunity to create multiple solutions for problems instead of looking for only one right answer. The use of LEGO® bricks also provides fact practice and recall that is much more entertaining and effective than the old-fashioned rote memorizing technique.

LEGO® bricks are great tools for bringing many mathematical concepts to life: basic cardinality and counting, addition and subtraction, multiplication and division, fractions, data and measurement, and statistics and probability. Using LEGO® bricks fosters discussion, modeling, collaboration, and problem solving. These are the 21st century skills that will help our students learn and be globally competitive.

The use of a common child's toy to do math provides a universal language for math. Children everywhere recognize this manipulative. It's fun to learn when you're using LEGO® bricks!

USING A BRICK MATH JOURNAL

Journaling in math is an exciting way for students, teachers, and parents to review and share what is going on in the math classroom. A math journal is a resource that students can use for years to practice and review math concepts.

I recommend having your students start a Brick Math journal when you begin using LEGO® bricks to help teach math concepts. Here's how to use a Brick Math journal with the activities in this book: In each chapter, students begin in Part 1 (Show Them How) by building models that are teacher-directed. In Part 2 (Show What You Know), students build their own LEGO® brick models in response to specific prompts. Finally, students draw their LEGO® brick models on paper in their own Brick Math journals. The journal serves as a record of the physical models built that students can refer to over and over. The Brick Math journal can also serve as a form of assessment for teachers, a source for conferences, and as a way to identify if a student has any misconceptions in learning the topic.

Use these steps to create a Brick Math journal:

1. Use a composition book for each student. Set up pages one and two as the table of contents.

2. Have students number each page on the top outside corner.

3. Photocopy the base plate paper in the Appendix. Students will use the base plate paper to record their solutions, drawings, reflections, etc. Students will glue the base plate paper onto a journal page after they have drawn and colored their solutions.

4. When students glue in their drawings, they should label them with the title of the activity and then make the entry in the journal's table of contents.

FINDING FACTORS

Students will learn/discover:
- The meaning of the term "factors"
- How to find all the factors of numbers
- How to make models of factor families
- Vocabulary:
  - **Factors:** Factors are numbers you can multiply together to get another number. Example: 2 and 3 are factors of 6; 2 and 4 are factors of 8.

Why is this important?
Students need to be able to identify all the factors of numbers before they can work on Least Common Multiples and Greatest Common Factors. Understanding the link between multiplication facts and division facts is crucial for students to prepare for upper levels of math, such as fractions. Knowing fact families and factors will help when learning to multiply larger numbers and will help with understanding division, which is often taught simultaneously with multiplication.

Brick Math journal:
After students build their models, have them draw the models on base plate paper and keep them in their Brick Math journals (see page 7 more about the Brick Math journal). Recording the models on paper after building with the LEGO® bricks helps reinforce the concepts.

Part 1: Show Them How
Model how to find all the factors of 16

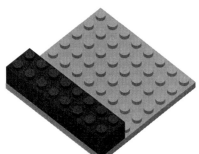

1. Place a 2x8 brick or a 1x16 brick on a base plate.

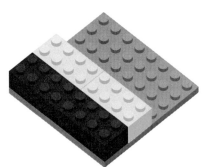

2. Place two bricks that are the same and, when placed next to the 16-stud brick, are equivalent in size and show two halves of the 16-stud brick. Use two 2x4 bricks or two 1x8 bricks.

3. Ask students: Can you find three bricks of equal size equivalent to the size of the 16-stud brick?

Let students look and think, and discover that the answer is no.

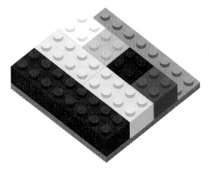

4. Ask students: Can you find four bricks of equal size equivalent to the size of the 16-stud brick?

Let students look and think, and discover that the answer is four 2x2 bricks or four 1x4 bricks.

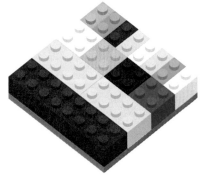

5. Ask students: Can you find the next number of equal-sized bricks that are equivalent to the size of the 16-stud brick?

Let students discover that five, six, and seven bricks don't work. Let them discover that the answer is eight 1x2 bricks.

6. Ask students: Can you find the next number of equal-sized bricks that are equivalent to the size of the 16-stud brick?

Let students discover that the answer is sixteen 1x1 bricks.

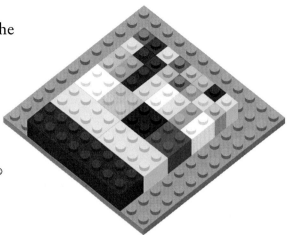

7. Name all the factors of 16 by looking at the LEGO® bricks on the base plate.

Answer: 16, 8, 4, 2, and 1.

Part 2: Show What You Know

1. Can you build a model to show all the factors of 6?

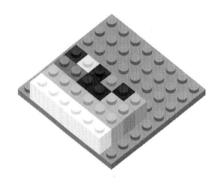

Solution A:
This model is a possible solution, showing factors 6, 3, 2, and 1.

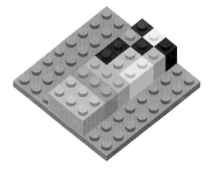

Solution B:
This model uses a different combination of bricks. Students who create this model could also explain that there are 2 sets of 3 in 6, and 3 sets of 2 in 6.

Show What You Know

2. Can you build a model to show all the factors of 8?

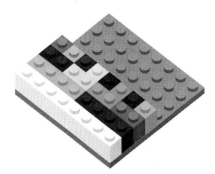

Solution A

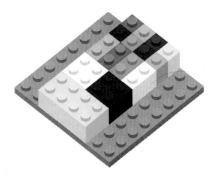

Solution B

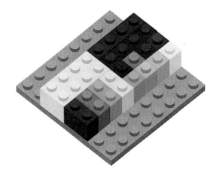

Solution C

Students who create models B or C could also explain that there are 2 sets of 4 in 8, and 4 sets of 2 in 8.

MULTIPLICATION USING SET MODELS

SUGGESTED BRICKS

Size	Number
1x1	10 (5 colors)
1x2	10 (5 colors)

Have a variety of bricks for these activities to provide opportunities for students to offer different solutions and for the teacher to create other models.

Note: A number of 1x10 or 1x12 bricks are also needed to serve as set separators.

Note: Using a base plate will help keep the bricks in a uniform line. One small and one large base plate is suggested for these activities.

Students will learn/discover:
- How to model multiplication as sets of numbers
- The structural design of multiplication problems
- The meanings for the numbers in multiplication problems
- Vocabulary:
  - **Set**
  - **Group**
  - **Multiplicand**
  - **Multiplier**
  - **Product**
  - **Factors**

Why is this important?
Students in grades 2 and 3 build on the sorting ideas learned in K – 1. They give names to sets and interpret information about sets, number of sets, number of groups, and "how many in all" as repeated addition. This activity is designed to help students make meaning of the term "set."

Brick Math journal:
After students build their models, have them draw the models on base plate paper and keep them in their Brick Math journals (see page 7 more about the Brick Math journal). Recording the models on paper after building with the LEGO® bricks helps reinforce the concepts.

Part 1: Show Them How #1
Model 4 sets of 6

1. Use the LEGO® bricks to make a model of 4 sets of 6. Make sure students are using 1x1 bricks because it is a model of sets of six ones.

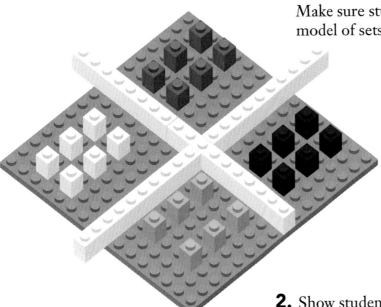

2. Show students that 4 sets of 6 is written as **4 x 6**

3. Have students use one-to-one correspondence (counting by ones, pointing to each stud) to determine the answer. Students should demonstrate knowledge that counting four sets of six in each set will result in a total of 24.

4. Make sure students understand that 4 equals the number of sets (groups) and 6 equals the number in each set. If they learn this structure for multiplication facts, they will understand the difference between 4 x 6 and 6 x 4.

Show Them How #2
Model 5 sets of 3

1. Use bricks to model 5 sets of 3.
Make sure students use 1x1 bricks.

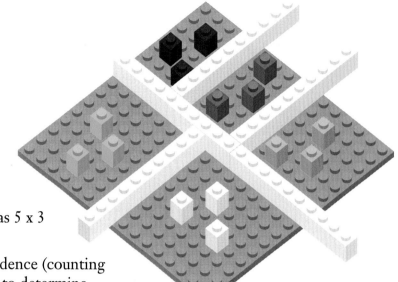

2. Show students that 5 sets of 3 is written as 5 x 3

3. Have students use one-to-one correspondence (counting by ones, pointing to each stud) if needed to determine the answer. This is a good time to utilize the skip-counting methods learned in grades K - 1 and count by threes. Students should demonstrate that these methods result in the answer to the problem of 15.

4. Make sure students understand that the 5 equals the number of sets (groups) and the 3 equals the number in each set. If they learn this structure for multiplication facts, they will understand the difference between 5 x 3 and 3 x 5.

Introduce the terms "multiplier" (the number of groups or sets) and "multiplicand" (the size of each group). In this problem, the multiplier is 5 and the multiplicand is 3.

Introduce the term "factors" as the name for the numbers 5 and 3.

5. Ask students to explain their models, and have them write an explanation and draw their models in their Brick Math journals.

Part 2: Show What You Know #1

1. Can you make a model of 2 sets of 6? Explain your model.

Answer: This model shows two sets with six 1x1 bricks in each set.

2. Can you use math vocabulary to talk about the numbers that your model includes? Can you write an equation for your model?

Answer: The multiplicand is 6 because there are 6 in each set, and the multiplier is 2 because there are 2 groups or sets.

The equation is 2 x 6 = 12.

The product is 12 since you can skip-count by 6s and get 12 on the second skip. (6,12)

Or: The product is 12, using one-to-one correspondence to count the individual bricks: there are 12 in all.

Show What You Know #2

1. Can you make a model of 8 sets of 3? Explain your model.

2. Can you use math vocabulary to talk about the numbers that your model includes? Can you write a problem for this equation that includes the product?

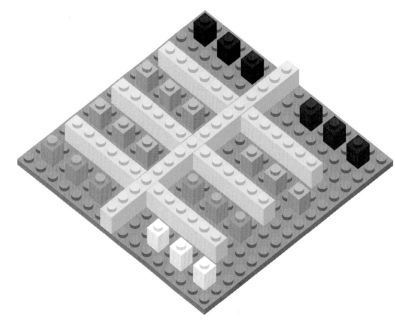

Answer: 8 x 3 = 24
There are several ways that students can model, count, and find the equation 8 x 3 = 24. They can skip-count by threes, add 12 and 12 for each side, count repeatedly by threes, or use one-to-one correspondence.

Students should include this information: The multiplicand is 3 because the set size is 3 and the multiplier is 8 because there are 8 groups or sets. The product is 24.

3

SUGGESTED BRICKS

Size	Number
1x1	30
1x2	16
1x3	8-10
1x4	8
1x6	8
1x8	4
1x10	2
1x12	4
1x16	1
2x2	10
2x3	6-8
2x4	6-8

Note: Using a base plate will help keep the bricks in a uniform line. One large base plate is suggested for these activities.

FACT FAMILIES

Students will learn/discover:
- To model all the facts within one fact family

Why is this important?
Knowing the fact families will help students learn basic multiplication facts and apply them in everyday mental math.

Note: Since there are no 5-stud bricks in the basic LEGO® bricks set, do not have students model facts that are products of numbers divisible by 5. The same applies to numbers divisible by 7 and 9. If you use LEGO® Technic bricks, you can expand your modeling.

Brick Math journal:
After students build their models, have them draw the models on base plate paper and keep them in their Brick Math journals (see page 7 more about the Brick Math journal). Recording the models on paper after building with the LEGO® bricks helps reinforce the concepts.

Part 1: Show Them How #1

1. Use bricks to model the fact family for 6.

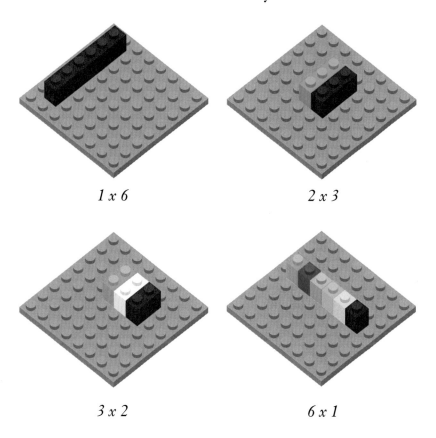

1 x 6   *2 x 3*

3 x 2   *6 x 1*

2. Write all the fact families for 6 in your Brick Math journal. Draw your models.

Students should write: 1 x 6, 2 x 3, 3 x 2, and 6 x 1.

Show Them How #2

1. Model the fact family for 8.

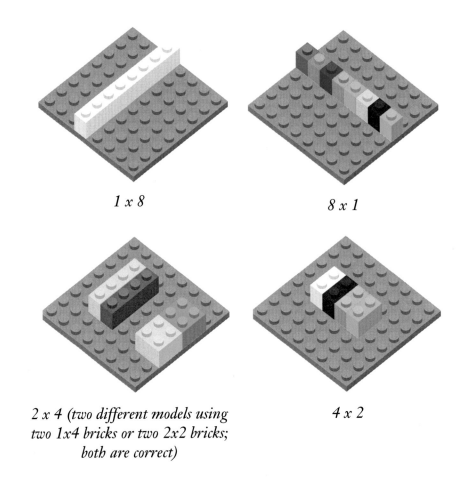

1 x 8

8 x 1

2 x 4 (two different models using two 1x4 bricks or two 2x2 bricks; both are correct)

4 x 2

2. Write all the fact families for 8 in your Brick Math journal. Draw your models.

Students should write: 4 x 2, 2 x 4, 1 x 8, and 8 x 1.

Part 2: Show What You Know #1

1. Can you model the fact family for 12?

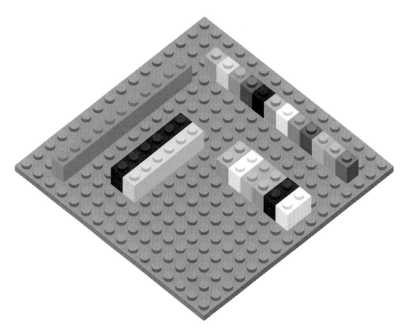

1 x 12, 12 x 1, 2 x 6, and 6 x 2

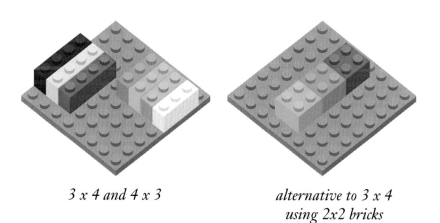

3 x 4 and 4 x 3　　　　*alternative to 3 x 4 using 2x2 bricks*

2. Write all the fact families for 12 in your Brick Math journal. Draw your models.

Answer: 1 x 12, 12 x 1, 2 x 6, 6 x 2, 3 x 4, and 4 x 3.

Show What You Know #2

1. Can you model the fact family for 16?

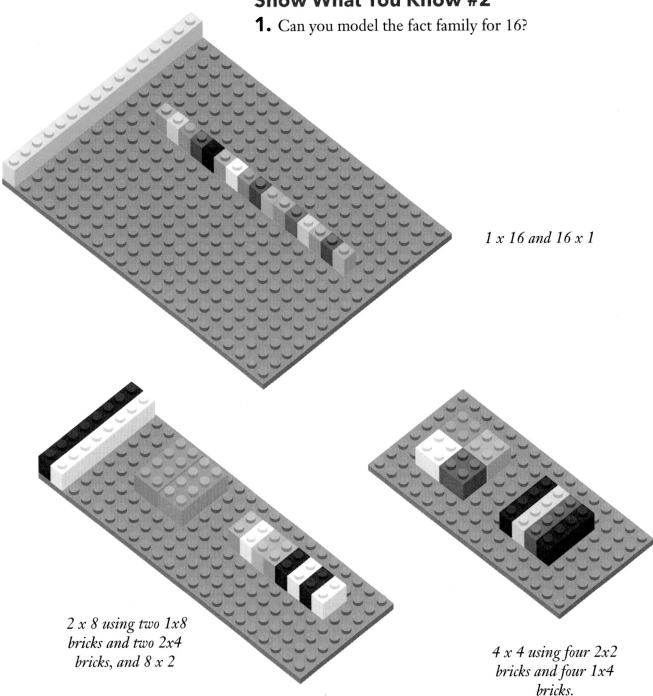

1 x 16 and 16 x 1

2 x 8 using two 1x8 bricks and two 2x4 bricks, and 8 x 2

4 x 4 using four 2x2 bricks and four 1x4 bricks.

2. Write all the fact families for 16 in your Brick Math journal. Draw your models.

Answer: 1 x 16, 16 x 1, 2 x 8, 8 x 2, and 4 x 4.

Show What You Know #3

1. Can you model the fact family for 24?

Note: There is not one brick to use as a model of 1 x 24. Use bricks that are all of one color whose studs add up to 24.

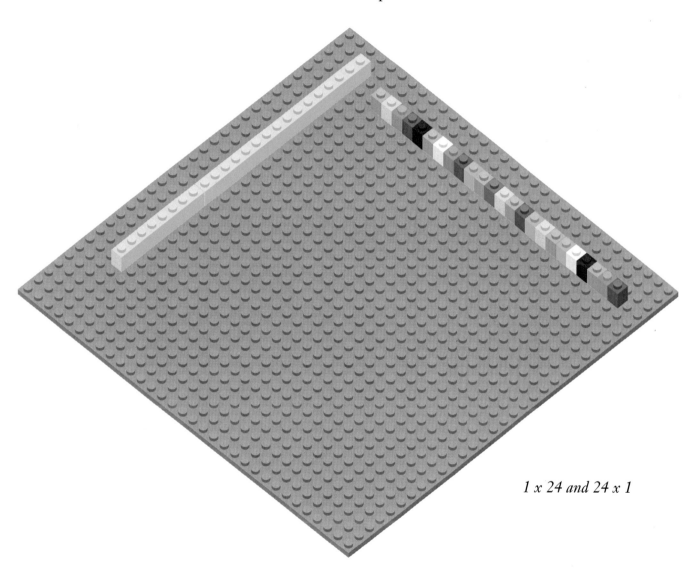

1 x 24 and 24 x 1

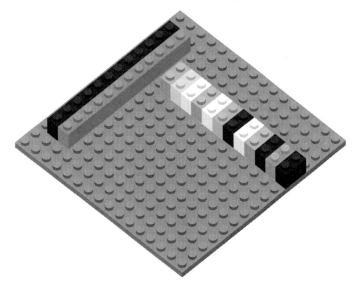

2 x 12 and 12 x 2

3 x 8 using three 2x4 bricks and three 1x8 bricks, and 8 x 3 using eight 1x3 bricks.

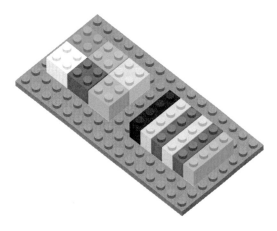

6 x 4 using six 2x2 bricks and six 1x4 bricks

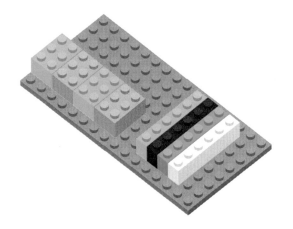

4 x 6 as four 2x3 bricks and four 1x6 bricks

2. Write all the fact families for 24 in your Brick Math journal. Draw your model.

Answer: 1 x 24, 24 x 1, 2 x 12, 12 x 2, 3 x 8, 8 x 3, 4 x 6, and 6 x 4.

BLOCKS AND BRICKS GAME

This activity creates a game-like learning environment to help students learn multiplication facts. Students play together in pairs to create models of multiplication facts.

Students will learn/discover:
- Basic multiplication facts through 6 x 6
- Vocabulary:
  - **Multiplier:** The number telling how many sets, or the number that tells how many times to multiply. For example, 2 x 4 means 2 sets of 4, so 2 is the multiplier.
  - **Multiplicand:** The number being multiplied. For example, 2 x 4 means 2 sets of 4, so 4 is the multiplicand.
  - **Frame of Reference:** The term that tells the item/contextual reference being multiplied, such as people, animals, studs, etc.

Why is this important?
Knowing basic facts leads to a greater ability to do mental math when math becomes more difficult. The game also prepares students to apply math in real contextual situations.

Brick Math journal:
After students build their models, have them draw the models on base plate paper and keep them in their Brick Math journals (see page 7 for more about the Brick Math journal). Recording the models on paper after building with the LEGO® bricks helps reinforce the concepts.

SUGGESTED BRICKS

Size	Number
1x1	20
1x2	10
1x3	10
1x4	10
1x6	3

Note: These suggested bricks are the minimum for each pair of game players.

Note: A number of 1x10 or 1x12 bricks are also needed to serve as set separators.

Note: Using a base plate will help keep the bricks in a uniform line. One large base plate is suggested for these activities.

Note: One set of two regular number cubes (dice) is needed for each pair of players.

Note: If the number 5 is rolled, model it by combining one 1x2 brick and one 1x3 brick, or by combining one 1x1 brick and one 1x4 brick.

Part 1: Show Them How

1. Have students make this table in their Brick Math journals:

Multiplcand	Multiplier	Final Model Sketch	Problem/Solution

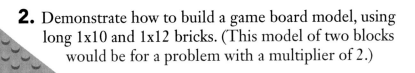

2. Demonstrate how to build a game board model, using long 1x10 and 1x12 bricks. (This model of two blocks would be for a problem with a multiplier of 2.)

3. Students play the game in pairs. Each player rolls the number cubes to create his/her own multiplication problem to model and solve.

Player 1 rolls one number cube. This number is the multiplier. (For this example, use 2 and 3 as the two digits rolled.)

4. Player 1 builds a game board with two blocks, modeling the multiplier of 2.

5. Player 1 models the multiplicand of 3 with three 1x1 bricks or one 1x3 brick in each block.

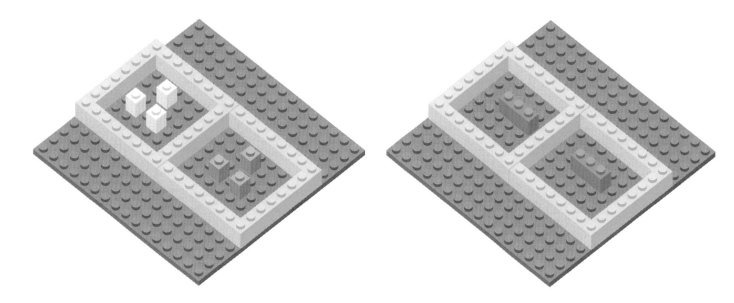

6. Player 1 counts the total number of studs in the model to find the solution to the problem (6).

7. At the same time, player 2 does the same procedure as player 1, creating his/her own multiplication problem to solve and modeling the solution.

8. Both players compare their models and discuss their solutions. In the table in their Brick Math journals, each player writes his/her problem and the solution.

Multiplier	Multiplicand	Final Model Sketch	Problem/Solution
2	3		2 x 3 = 6 studs

Make sure students use the word "studs" as a frame of reference that describes exactly what item is being multiplied. Practice the game one more time to be sure everyone knows how to play.

For a second example, assume that 4 and 2 have been rolled.

1. Ask students questions before they start to build models.

Ask: How many blocks are needed to show the multiplier?
Answer: 4 blocks

Ask: What bricks are needed to show the multiplicand?
Answer: Bricks with 2 studs in each block show the multiplicand of 2.

2. Have students build the game board with 4 blocks, then
fill each block with either two 1x1 bricks in each block
or one 1x2 brick in each block.

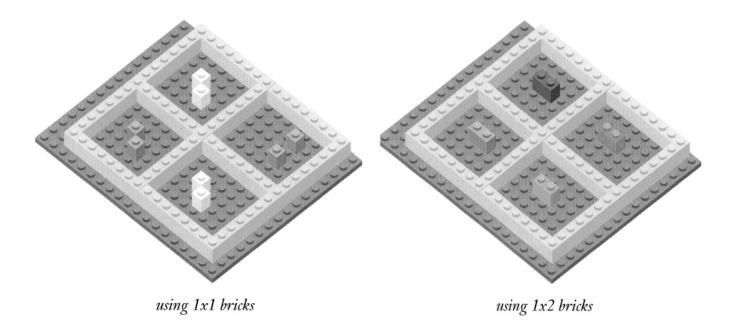

using 1x1 bricks   *using 1x2 bricks*

3. Have students write the problem and its solution in
their Brick Math journals:

4 x 2 = 8 studs

Part 2: Show What You Know
Play the game!

Directions:

1. Pair students to play the game.

2. Each player rolls one die to determine the multiplier for his/her problem. Each player builds a game board with the number of blocks to model his/her multiplier.

3. Each player rolls one die to determine the multiplicand for his/her problem. Each player fills his/her game board blocks with the appropriate bricks to model that multiplicand.

4. Both players compare their models and discuss their solutions.

5. In the table in their Brick Math journals, each player writes his/her problem and the solution and makes a sketch of the model.

6. After the agreed-on number of rounds, players add up the total number of studs in the solution column. The player with the highest number wins the game!

5

SUGGESTED BRICKS

Size	Number
1x1	10
1x2	25
	(various colors)
1x3	15
1x10	5

Note: A number of 1x10 or 1x12 bricks are also needed to serve as set separators.

Note: Using a base plate will help keep the bricks in a uniform line. One small and one large base plate is suggested for these activities.

MULTIPLICATION USING PLACE VALUE/BUNDLING MODELS

Students will learn/discover:
• How to model multiplication based on place value bundles

Why is this important?
When learning to multiply numbers greater than 9, modeling the multiplication using the set model becomes unwieldy. Using bricks to represent place values or bundles of 10, 100, 1000, etc., is an efficient way to model. It also helps expand students' understanding of place value as it pertains to multiplication.

This method of modeling helps leads to the understanding of multiplication through partial products.

Brick Math journal:
After students build their models, have them draw the models on base plate paper and keep them in their Brick Math journals (see page 7 for more about the Brick Math journal). Recording the models on paper after building with the LEGO® bricks helps reinforce the concepts.

Part 1: Show Them How

1. Model 2 x 25. What does that mean?

Answer: 2 sets of 25

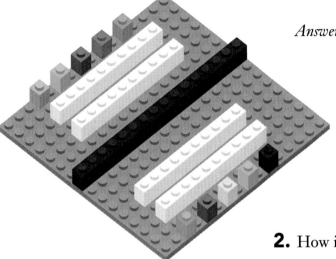

2. How is this different from 25 x 2?

Answer: That means 25 sets of 2.

Model 25 x 2.

Answer: (Make sure students are modelling correctly here.)

In the 2 x 25 model, the tens and ones are bundled together into two tens and 5 ones. In the 25 x 2 model, there are 25 separate sets of twos.

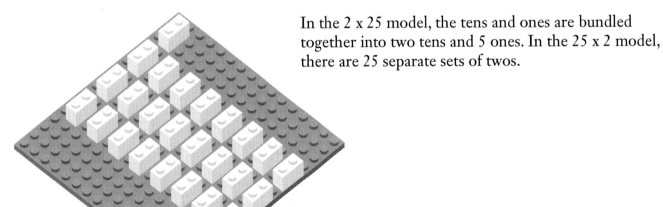

3. Model the product of the problem using the bricks in the 2 x 25 model.

Answer: Bring together the two sets of 25, as shown in the place value model, into one number representation. Bundle the four 1x10 bricks together and the ten 1x1 bricks together.

Regroup the ten ones into one ten. Now the model shows 5 tens, or 50.

Part 2: Show What You Know #1

1. Can you make a place value model of 3 x 12? What does the 3 represent? What does the 12 represent?

Answer: The 3 represents three sets and 12 represents that there are twelve in each of the 3 sets.

2. How is that problem different from 12 x 3? Can you show the difference? Can you explain the difference?

Answer: 12 x 3 shows 12 separate sets with 3 in each set.

The problems are different because the number of sets is different and the number in each one is different.

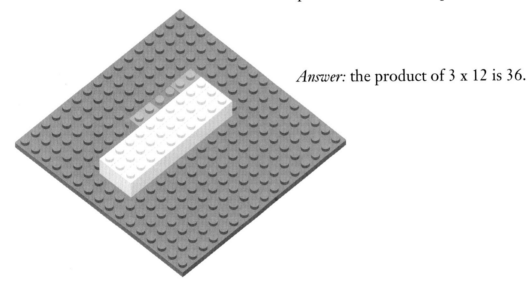

3. Can you bundle the 3 x 12 model together to show the product? What is the product?

Answer: the product of 3 x 12 is 36.

Show What You Know #2

1. Can you use a place value/bundle model to show 3 sets of 13 or 3 x 13?

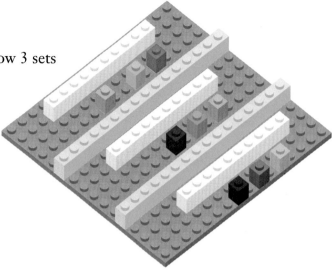

2. How does this look different than 13 x 3?

Answer: The model of 13 x 3 uses thirteen 1x3 bricks to show 13 groups of 3.

3. What can you say about the product of 3 x 13 and 13 x 3?

Answer: The product is 39. The product is the same for both problems. The model of 3 x 13 shows place value and decomposition, which is the breaking apart of values into smaller numbers. The model of 13 x 3 shows sets.

4. Can you model the product of 3 x 13 with a place value/bundle model?

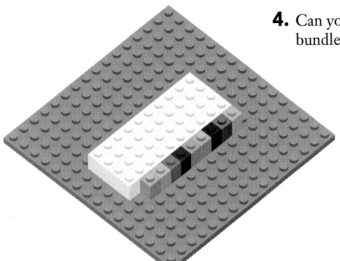

Show What You Know #3

1. Can you show the difference between 2 x 24 and 24 x 2?

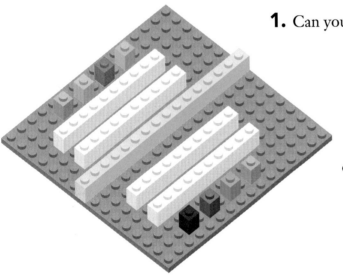

Answer: For 2 x 24, the model shows two groups of 2 tens and 4 ones.

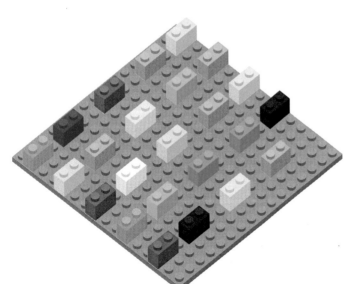

Answer: For 24 x 2, the model shows 24 sets of 2 ones.

3. What is the product of the two problems?

Answer: 2 x 24 = 48 and 24 x 2 = 48

The products of these two models are the same but they mean different things.

4. Can you model the product using a place value/bundle model?

More to discover:
Try more problems:

3 x 14

4 x 22

6

MULTIPLICATION USING ARRAY/AREA MODELS

Students will learn:
- How to model multiplication using arrays

Why is this important?
The array/area model helps students understand one-digit multiplication.

Brick Math journal:
After students build their models, have them draw the models on base plate paper and keep them in their Brick Math journals (see page 7 for more about the Brick Math journal). Recording the models on paper after building with the LEGO® bricks helps reinforce the concepts.

Part 1: Show Them How #1

1. Make an array/area model of 3 x 2 using one brick.

    Array model showing 3 x 2: 3 studs across and 2 studs down.

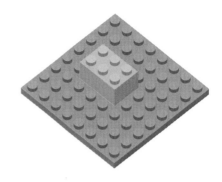

2. Make an array/area model of 2 x 3 using one brick.

    Array model showing 2 x 3: 2 studs across and 3 studs down.

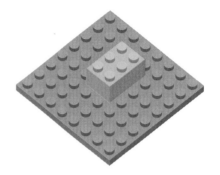

3. By counting the studs, students can determine the answer to each problem: 6.

4. Use the terminology:

    The **product** (6) is found by multiplying the two **factors** (3 and 2) together: 3 x 2 = 6

5. Discuss how these two array models are alike and different.

    Both give the same answer, but the solution is different because the orientation (vertical versus horizontal) is different. These two multiplication facts (2 x 3 and 3 x 2) have the same outcome in terms of answer or product. However, they mean something different in terms of geometric concepts. For example, if you are building a store and want the front of the store to have the longest side facing the street front, you want to use the 3 x 2 model. For engineering, orientation is important when modeling multiplication!

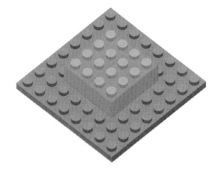

Show Them How #2

1. Model 4 x 4 using two bricks.

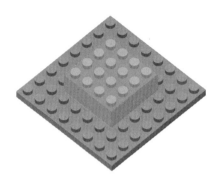

2. Ask students what they notice about the orientation in this model that is different from the previous one.

Answer: The orientation does not matter because all sides are the same length.

3. Make another array model that this holds true for.

Answer: Building any square number will result in this effect.

Examples:

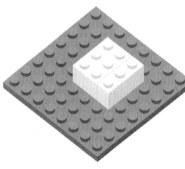
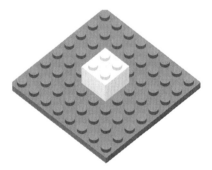

3 x 3 = 9   *2 x 2 = 4*

Part 2: Show What You Know #1

1. Can you model an array of 4 x 6 using 3 bricks?

Explain your model in your journal. Draw your model. What is the product?

Answer: This model shows 4 across and 6 down (4 x 6). Counting the studs gives the answer of 24. The product of multiplying the two factors is written as 4 x 6 = 24.

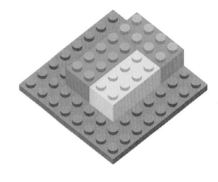

2. Does this look different from 6 x 4? Explain your thinking.

Answer: The 6 x 4 model has the longer side horizontal with 6 studs as the length.

Show What You Know #2

1. Can you model an array of 3 x 6? Explain the model. What is the product?

Answer: This model shows 3 studs horizontally and 6 studs vertically. The equation is written as 3 x 6 and the product is 18.

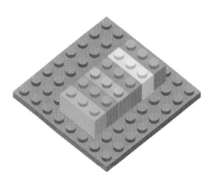

2. Can you model and explain how the orientation changes this model?

Answer: This model shows 6 studs across (horizontally) and 3 studs down (vertically). The problem is 6 x 3 = 18.

The answers are the same but the models are different.

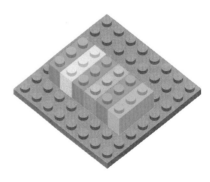

SUGGESTED BRICKS

Size	Number
1x1	30
1x2	10
1x3	10
1x4	5
1x10	5
2x4	6-8

Note: Have a variety of colors of all bricks. Use additional 1x10, 1x12, and 1x16 bricks as set dividers.

Note: Using a base plate will help keep the bricks in a uniform line. Use one large base plate for these activities.

MULTIPLICATION MODELING CHALLENGE

This activity brings together the knowledge gained from the previous chapters. It can serve as an assessment task to be sure students have learned how to model using sets, place value, and arrays. It can also be set up as a station task in the classroom.

Brick Math journal:
After students build their models, have them draw the models on base plate paper and keep them in their Brick Math journals (see page 7 for more about the Brick Math journal). Recording the models on paper after building with the LEGO® bricks helps reinforce the concepts.

1. Can you model 4 x 5 using a set model?

Explain your model in your Brick Math journal. Draw your solution. What is the product?

How is 4 x 5 different from 5 x 4? Can you show the difference?

Answer: Students should point out that the two models have the same product (20) but that one has 4 sets of 5 and one has 5 sets of 4, making the number of groups and number in each group different.

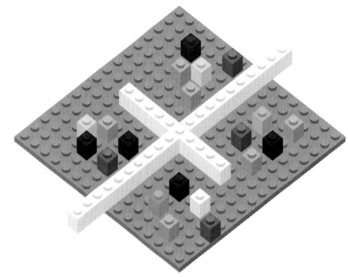

4 sets of 5

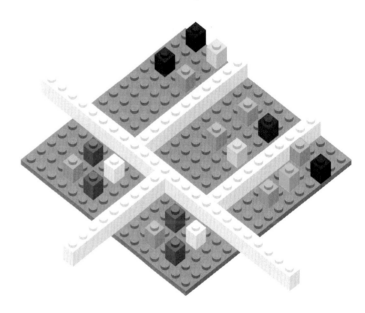

5 sets of 4

2. Can you model 3 x 14 using a place value model? Explain your model in your Brick Math journal. Draw your solution. What is the product?

Answer: Students should be able to combine tens (3) and combine ones (12) to show an answer of 30 + 12 = 42.

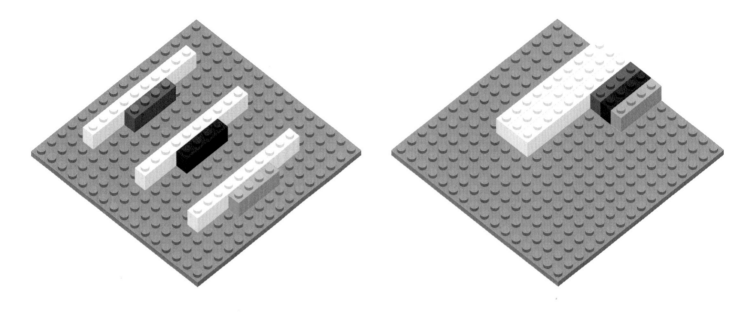

More for students to discover:
This is a great time to discuss expanded form and to discuss regrouping with students. Model how 12 in this answer is equal to 1 set of ten and 2 ones.

Answer: Move the 12 ones to line up next to a 1x10 brick. Two 1x1 bricks are left over. Students can see that there is another set of ten in the answer (4 sets of ten) and 2 more 1x1 bricks, making the answer 42.

3. Can you make an array/area model of 4 x 8?

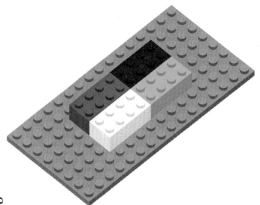

4. Can you make an array/area model of a square number?

Answer: Various models apply, including any that result in a model that is a square (2 x 2, 3 x 3, 4 x 4, etc.).

8

SUGGESTED BRICKS

Size	Number
1x1	25
1x2	30
1x3	1
1x4	1

Note: Using a base plate will help keep the bricks in a uniform line. Two base plates are suggested for these activities.

Note: It is helpful if sets are represented by different colors of bricks and the set multiplier bricks are all the same color within the problem.

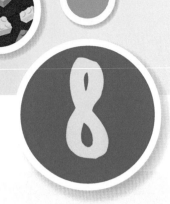

MULTIPLYING TWO-DIGIT NUMBERS BY ONE-DIGIT NUMBERS

Students will learn/discover:

- How to multiply two-digit numbers by one-digit numbers
- The role of place value in reading and understanding numbers
- The use of expanded form when writing a number by its place value
- That multiplication is repeated addition
- Vocabulary:
    - **Multiplier:** the number that determines how many sets, or how many times a number is to repeated or multiplied
    - **Multiplicand:** the number that is multiplied or repeated

Why is this important?

It is important that students understand the value of digits within a number prior to multiplication in order to determine reasonability of the solution in the end. For example, if a student tries to multiply 23 x 6 and does not understand that the digit "2" represents tens (thus equivalent to 20), he/she may arrive at an incorrect solution.

This activity is intended for those just beginning to multiply with two digits. It also reinforces the concept that multiplication is repeated addition.

Brick Math journal:

After students build their models, have them draw the models on base plate paper and keep them in their Brick Math journals (see page 7 for more about the Brick Math journal). Recording the models on paper after building with the LEGO® bricks helps reinforce the concepts.

Part 1: Show Them How #1
Multiplying two digits by one digit

Make sure students understand this modeling technique before moving on to solve the multiplication problem.

Model the number 1,111 as shown and explain:

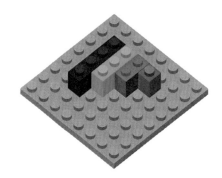

- The 1x1 brick represents ones (in this case, 1)
- The 1x2 brick represents tens (in this case, 10)
- The 1x3 brick represents hundreds (in this case, 100)
- The 1x4 brick represents thousands (in this case, 1000)
- The expanded form of this number is 1000 + 100 + 10 + 1

After students understand this expanded form modeling technique, move on to the multiplication solution process.

Solve the problem 2 x 10 using the expanded form modeling technique.

1. Ask students what this problem represents.

Two sets of 10.

2. Ask students which number is the multiplier and which number is the multiplicand.

The multiplier is 2 and the multiplicand is 10.

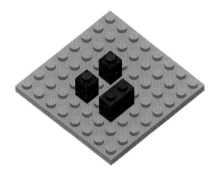

3. Model the problem 2 x 10.

To model the muliplicand of 10, use one 1x2 brick (representing the 10). To model the multiplier of 2, use two 1x1 bricks (representing 2 sets).

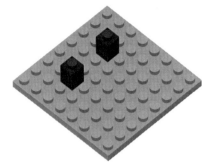

4. Take each 1x1 brick and decompose them into two sets on another base plate.

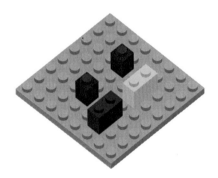

5. Take one 1x2 brick (representing 1 ten) and place it next to the first set marker to show one set of 10. Take another 1x2 brick and place it next to the other set marker.

6. Remove the set marker bricks (the 1x1 bricks).

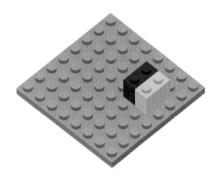

7. Place the 1x2 bricks together to represent 2 tens, or 20. The solution is 20.

Point out that this model also shows 10 + 10 = 20, using repeated addition.

Show Them How #2
Solve 2 x 23

1. Ask students what this problem represents.

Two sets of 23.

2. Ask students which number is the multiplier and which number is the multiplicand.

The multiplier is 2 and the multiplicand is 23.

3. Model the problem 2 x 23.

To model the multiplicand of 23, use two 1x2 bricks to represent 2 tens, or 20, and three 1x1 bricks to represent 3 ones. To model the multiplier of 2, use two 1x1 bricks to represent 2 sets.

4. Move the two 1x1 bricks to show how many sets are needed.

5. To show the multiplication process, first move two 1x2 bricks (representing 2 tens, or 20) next to each set marker.

6. Move three 1x1 bricks next to the first set, and then move three more 1x1 bricks next to the second set.

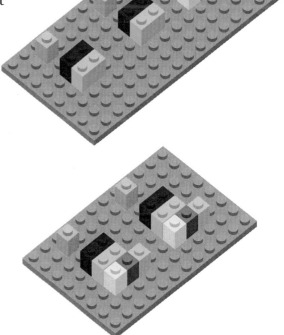

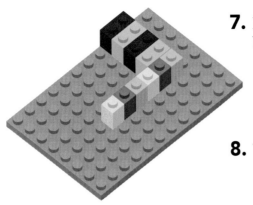

7. Remove the set markers from the model and group the bricks together to show the solution.

8. Write the expanded form of the problem.

10 + 10 + 10 + 10 + 3 + 3 = 40 + 6 = 46

Show Them How #3
Solve 3 x 34

1. Ask students what this problem represents.

Three sets of 34.

2. Ask students which number is the multiplier and which is the multiplicand.

The multiplier is 3 and the multiplicand is 34.

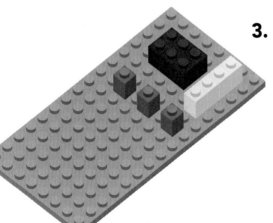

3. Model the problem 3 x 34.

Answer: To model the multiplicand of 34, use three 1x2 bricks to represent 3 tens, or 30, and four 1x1 bricks to represent 4 ones. To model the multiplier of 3, use three 1x1 bricks to represent 3 sets.

4. Move the three 1x1 bricks to show how many sets are needed.

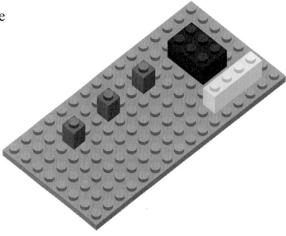

5. Show the multiplication process: first, move three 1x2 bricks (representing 3 tens, or 30) next to the first set marker, then move three more 1x2 bricks next to the second set marker, and three more 1x2 bricks next to the third set marker.

6. Do the same with the four 1x1 bricks (representing 4 ones). Move four 1x1 bricks next to the first set marker, then move four more 1x1 bricks next to the second set marker, and four more 1x1 bricks next to the third set marker.

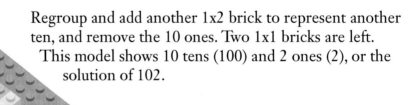

7. Remove the set markers from the model and group the bricks together to show the solution.

8. Write the expanded form of the problem.

$$30 + 30 + 30 + 4 + 4 + 4 = 90 + 12 = 102$$

More to discover:

Decompose the ones and regroup the tens.

The 1x10 brick next to the 1x1 bricks shows that there are more than ten ones.

Regroup and add another 1x2 brick to represent another ten, and remove the 10 ones. Two 1x1 bricks are left. This model shows 10 tens (100) and 2 ones (2), or the solution of 102.

Part 2: Show What You Know #1
2 x 22

Have students complete the model and use the terms: *sets of*, *multiplier*, and *multiplicand*. When completed, have students draw the models and explain their process in their Brick Math journals.

1. Model the problem and show the parts of the problem.

Answer: The model shows 2 tens and 2 ones multiplied by 2. There are 2 sets of 22. The multiplier is 2 and the multiplicand is 22.

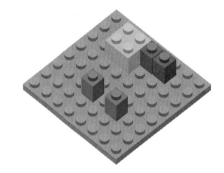

2. Show two sets and the process of multiplication. First, move the tens to next to the set markers.

Answer: Two sets of ten, or 20 + 20 = 40.

3. Next, show the ones added to the tens in each set.

Answer: Two ones added to each set of 20.

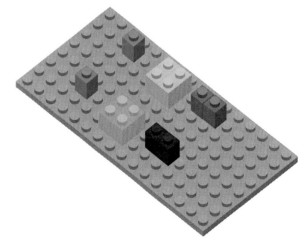

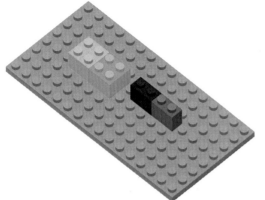

4. Remove the set markers and combine the place values.

Answer: The model shows the final solution of 4 tens and 4 ones, or, in expanded form, $20 + 20 + 2 + 2 = 40 + 4 = 44$.

Show What You Know #2
3 x 32

Students should identify 32 as the multiplicand and 3 as the multiplier.

Students should show each step of the multiplication process.

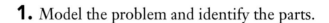

1. Model the problem and identify the parts.

Answer: To model the multiplicand of 32, use three 1x2 bricks to represent three tens or 30, and use two 1x1 bricks to represent 2 ones.

To model the multiplier of 3, use three 1x1 bricks.

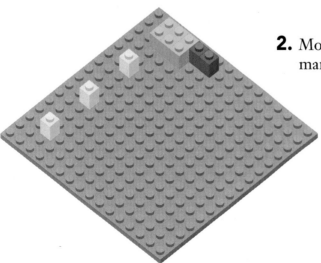

2. Move the three multiplier bricks to show the set markers.

3. Show the process defining each place value and the expanded form in the process.

Answer: Form 3 sets of 30 with 1x2 bricks, then add 3 sets of 2 ones.

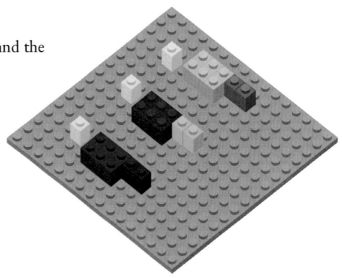

4. Remove the set markers and bring the model together to show 9 tens and 6 ones.

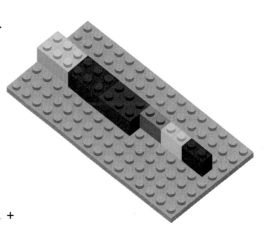

5. Write the expanded form.

Answer: 10 + 10 + 10 + 10 + 10 + 10 + 10 + 10 + 10 + 1 + 1 + 1 + 1 + 1 + 1 = 90 + 6 = 96

Show What You Know #3
4 x 15

Students should identify 15 as the multiplicand and 4 as the multiplier.

Students should show each step of the multiplication process.

1. Model the problem and identify the parts.

Answer: To model the multiplicand of 15, use one 1x2 brick to represent 1 ten or 10, and use five 1x1 bricks to represent 5 ones.

To model the multiplier of 4, use four 1x1 bricks.

2. Move the four multiplier bricks to show the set markers.

3. Show the process defining each place value and the expanded form in the process.

Answer: Form 4 sets of 10 with 1x2 bricks, then add 4 sets of 5 ones.

4. Remove the set markers and bring the model together to show 4 tens and 20 ones.

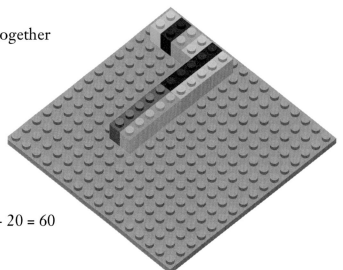

Write the expanded form.

Answer: 10 + 10 + 10 + 10 + 5 + 5 + 5 + 5 = 40 + 20 = 60

5. Regroup the ones to show two more tens.

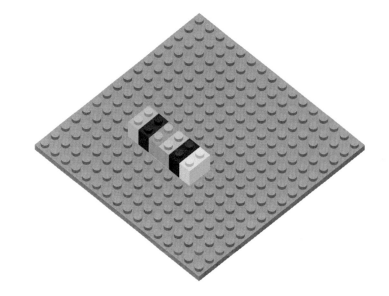

6. Write the expanded form.

Answer: 10 + 10 + 10 + 10 + 10 + 10 = 60

Challenge: Create another problem and have a partner show and explain the solution.

9

SUGGESTED BRICKS

Size	Number
1x1	25
2x1	30
3x1	10
4x1	10

Note: Using a base plate will help keep the bricks in a uniform line. Two base plates are suggested for these activities.

MULTIPLYING LARGER NUMBERS

Students will learn/discover:
- How to multiply larger numbers by single digit numbers
- The role of place value in reading and understanding numbers
- The use of expanded form when writing a number using its place value
- That multiplication is repeated addition
- Vocabulary:
  - **Multiplier:** the number of sets, or how many times a number is repeated or multiplied
  - **Multiplicand:** the number that is multiplied or repeated

Why is this important?
This activity provides practice with extending multiplication to larger numbers. This activity links expanded form from addition and subtraction to multiplication. Students must understand the value of digits within a number before multiplying to determine if the final solution is reasonable. Although the modeling process for this activity using bricks may be a bit cumbersome to learn, teachers find that students are very engaged in the process, which helps them understand the math more fluently. The math practices from the National Council of Teachers of Mathematics (NCTM) Principles and Standards encourage the development of fluent computing and practice with modeling processes in the elementary and middle grades.

Brick Math journal:
After students build their models, have them draw the models on base plate paper and keep them in their Brick Math journals (see page 7 for more about the Brick Math journal). Recording the models on paper after building with the LEGO® bricks helps reinforce the concepts.

Part 1: Show Them How #1
Solve 3 x 125

Make sure students understand the expanded form modeling technique used in Chapter 8 first.

1. Ask students what this problem represents.

Three sets of 125.

2. Ask students which number is the multiplier and which one is the multiplicand.

The multiplier is 3 and the multiplicand is 125.

3. Model the problem 3 x 125.

Answer: To model the multiplicand of 125, use one 1x3 brick (representing one hundred), two 1x2 bricks (representing 2 tens or 20), and five 1x1 bricks (representing 5 ones). To model the multiplier of 3, use three 1x1 bricks (representing 3 sets). It is helpful to use one color for these set marker bricks.

4. Take each 1x1 brick and break them into 3 sets on another base plate.

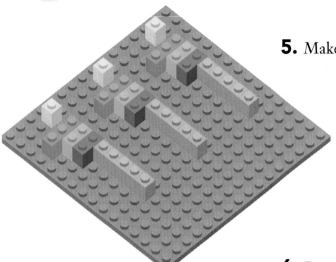

5. Make 3 sets of the multiplicand bricks.

6. Remove the set marker bricks (the 1x1 bricks).

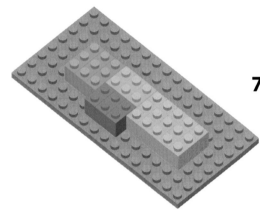

7. Place like blocks together to make hundreds, tens and ones.

8. Point out that this model shows the expanded form 300 + 60 + 15 in the solution.

Regroup the ones to one more ten (one 1x2 brick) and 5 ones, making the expanded form 300 + 70 + 5 = 375.

Show Them How #2
Solve 3 x 2,232

1. Model the problem.

Two 1x4 bricks are used to model the thousands (2,000).

2. Make 3 sets of the multiplicand of 2,232 based on the multiplier of 3.

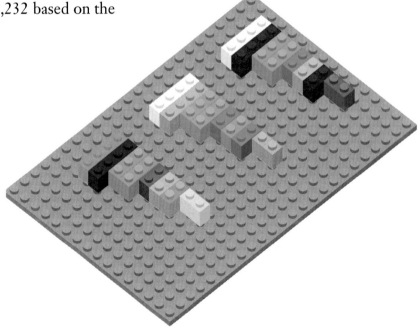

3. Place like blocks together to make thousands, hundreds, tens and ones.

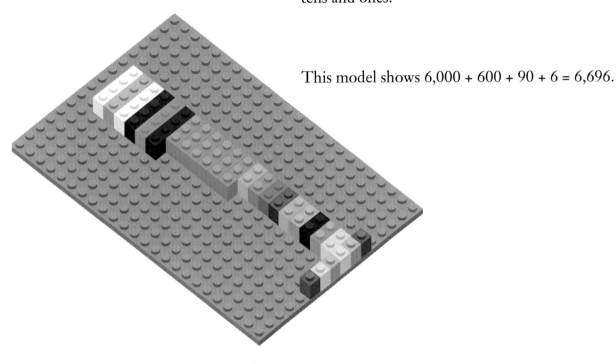

This model shows 6,000 + 600 + 90 + 6 = 6,696.

Part 2: Show What You Know #1
Solve 3 x 325

1. Model the problem and explain.

> *Answer:* The multiplicand of 325 is modeled with three 1x3 bricks, two 1x2 bricks, and five 1x1 bricks. Three set marker 1x1 bricks show the multiplier of 3.

2. Model the sets based on the multiplier of 3 and explain.

Answer: This model shows three sets of 325.

3. Place all the hundreds, tens, and ones together to show the expanded form of the solution and explain the solution.

Answer: This model shows
(9 x 100) + (6 x 10) + (15 x 1) = 975.

The expanded form can be written as:
900 + 60 + 15 or
900 + 70 + 5 if student decomposes the 15 into 1 ten and 5 ones.
The solution is 975.

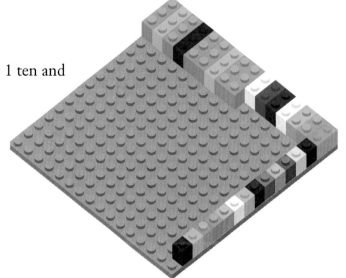

Show What You Know #2
Solve 4 x 1,323

1. Model and explain the problem.

> *Answer:* 1,323 is modeled with one 1x4 brick, three 1x3 bricks, two 1x2 bricks, and three 1x1 bricks. Four 1x1 bricks show the multiplier of 4.

2. Model the sets based on the multiplier of 4.

> *Answer:* This model shows four sets of 1,323.

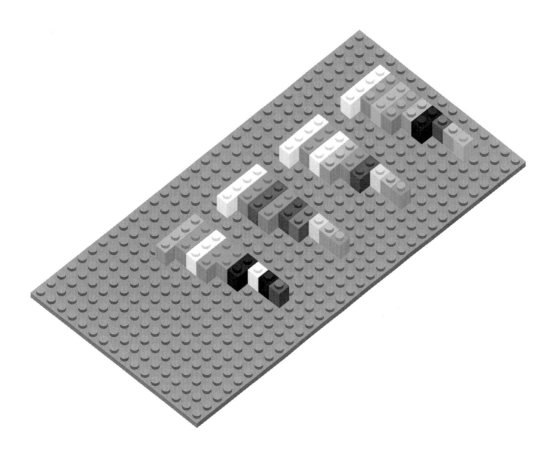

3. Place all the thousands, hundreds, tens, and ones together to show the expanded form of the solution and explain the solution.

Note: Expect students to be able to decompose and pull together sets of 10 and 100 in the final explanation.

Answer: This model shows
(4 x 1,000) + (12 x 100) + (8 x 10) + (12 x 1).

The expanded form can be written as
4,000 + 1,200 + 80 + 12.

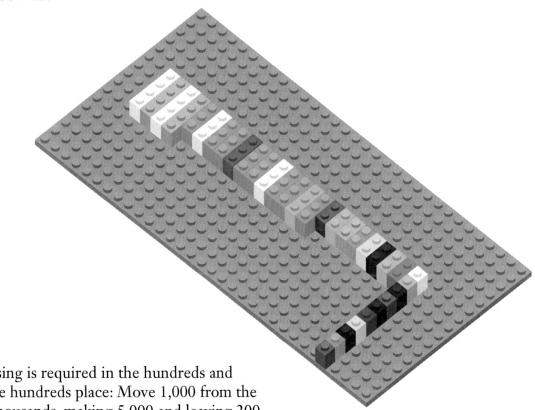

Some decomposing is required in the hundreds and the ones. For the hundreds place: Move 1,000 from the 1,200 into the thousands, making 5,000 and leaving 200.

For the ones place: move 10 into the tens, making 90 and leaving 2.

The problem in expanded form:

5,000 + 200 + 90 + 2 = 5,292.

Show What You Know #3
Solve 5 x 2,241

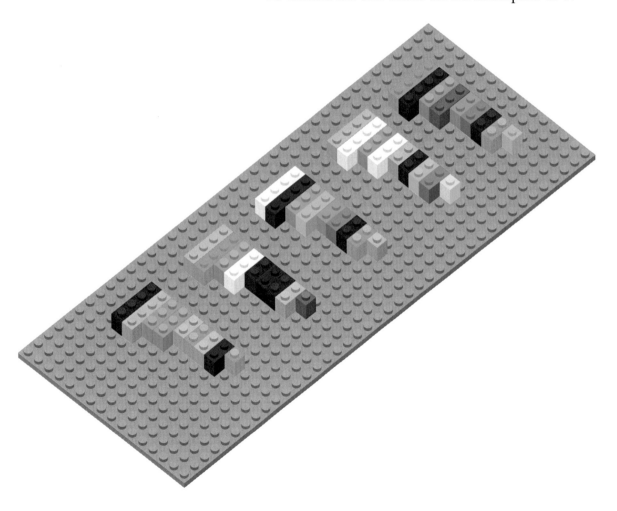

1. Model the problem and explain.

> *Answer:* The multiplicand of 2,241 is modeled with two 1x4 bricks, two 1x3 bricks, four 1x2 brick, and one 1x1 brick. Five 1x1 bricks show the multiplier of 5.

2. Model the sets based on the multiplier of 5.

3. Place all the thousands, hundreds, tens, and ones together to show the expanded form of the solution and explain the solution.

Note: Expect students to be able to decompose and pull together sets of 10 and 100 in the final explanation.

Answer: This model shows (10 x 1,000) + (10 x 100) + (20 x 10) + (5 x 1).

The expanded form of this number is 10,000 + 1,000 + 200 + 5 = 11,205.

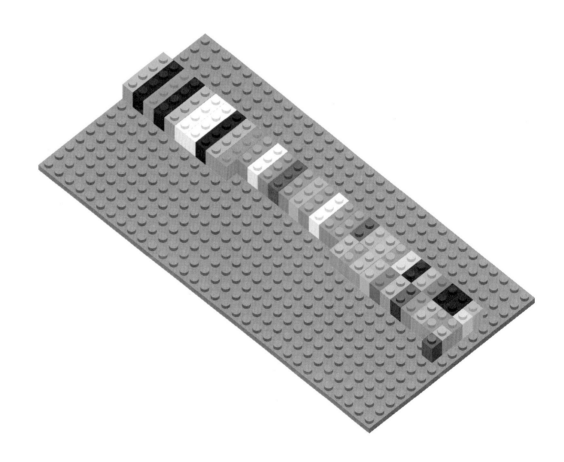

SUGGESTED BRICKS

Size	Number
1x1	30
1x2	20-30
1x3	10
1x4	10
1x6	10
1x10	4-8

Note: Using a base plate will help keep the bricks in a uniform line. One large base plate is suggested for these activities.

MULTIPLYING TWO-DIGIT NUMBERS BY TWO-DIGIT NUMBERS

Students will learn/discover:
* How to use both the place value and the array models to determine products when multiplying a two-digit number by another two-digit number

Why is it important?
When students learn to relate larger multiplication problems to place value it becomes easier to do mental multiplication. Being able to do mental math makes the application of math in everyday activities easier.

This technique gives students a visual connection between previously learned content and larger multiplication problems.

Brick Math journal:
After students build their models, have them draw the models on base plate paper and keep them in their Brick Math journals (see page 7 for more about the Brick Math journal). Recording the models on paper after building with the LEGO® bricks helps reinforce the concepts.

Part 1: Show Them How #1

Before introducing the method of modeling two-digit by two-digit multiplication, review the process of multiplying a two-digit number by a one-digit number, multiplying the ones by tens and the ones by ones. As an example, use 23 x 3:

23
x 3
breaks down into:
20 x 3 = 60 (tens x ones)
3 x 3 = 9 (ones x ones)
Product: 60 + 9 = 69

Using this idea, introduce the modeling method for multiplying a two-digit number by another two-digit number using a rectangular array model of bricks.

Model 14 x 12 using an array model

1. Think of 14 as 10 + 4 = 14 and think of 12 as 10 + 2 = 12.

Model 14 with one 1x10 brick and one 1x4 brick.

Model 12 with one 1x10 brick and one 1x2 brick and set it at a right angle to the model of 14. Overlap the ends of the 1x10 bricks.

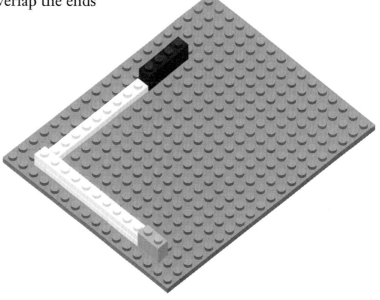

2. Set two 1x10 bricks at right angles to the other 1x10 bricks to make a rectangle, overlapping the corners. This rectangle represents tens x tens, and shows that 10 x 10 makes a rectangle of 100 studs. The 1x4 brick and the 1x2 brick should extend outside the rectangle.

3. Fill in the top with 1x4 bricks to the end of the rectangle. This represents the tens x ones, or 10 x 4 = 40.

4. Fill in the side with 1x2 bricks to the top of the rectangle. This represents the ones x tens, or 10 x 2 = 20.

5. The top right corner area needs to be filled to complete the rectangle. Fill that with eight 1x1 bricks. This represents the ones x ones, or 4 x 2 = 8.

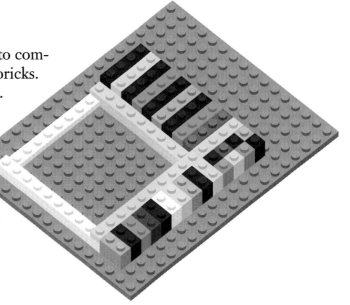

6. When this rectangle is completed, it shows the components of place value in an array model.

$$10 \times 10 = 100$$
$$10 \times 4 \ = \ \ 40$$
$$10 \times 2 = \ \ \ 20$$
$$2 \times 4 \ \ \ = \ \ \ \ 8$$

$$\overline{}$$

$$168$$

The product of 14 x 12 = 168.

Show Them How #2
Model the problem 16 x 11

1. Model the 16 as 10 + 6 and the 11 as 10 + 1.

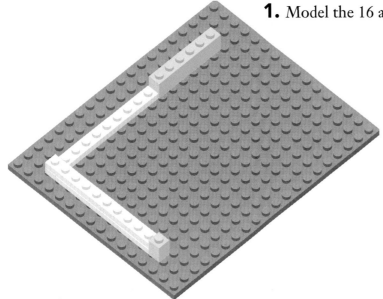

2. Make the rectangle to show 10 x 10 = 100.

The 1x6 and 1x1 bricks should extend outside the rectangle.

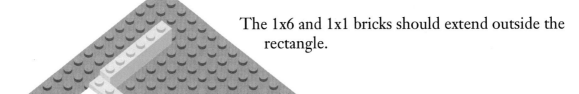

3. Fill in the top of the rectangle with 1x6 bricks to show the tens x ones, or 6 x 10 = 60. Fill in the right side of the rectangle with 1x1 bricks to represent the ones x tens, or 1 x 10 = 10.

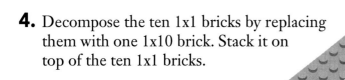

4. Decompose the ten 1x1 bricks by replacing them with one 1x10 brick. Stack it on top of the ten 1x1 bricks.

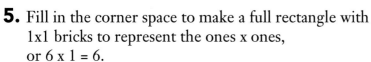

5. Fill in the corner space to make a full rectangle with 1x1 bricks to represent the ones x ones, or 6 x 1 = 6.

6. Add all the places together:

10 x 10 = 100
10 x 6 = 60
10 x 1 = 10
1 x 6   =   6
_____
       176

The product of 16 x 11 = 176.

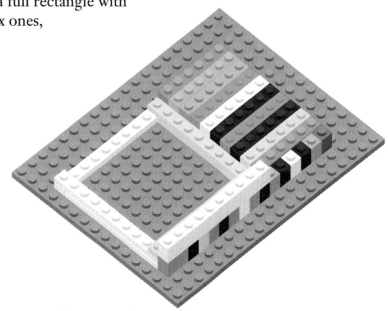

Part 2: Show What You Know

1. Can you model 13 x 12?

Students should:
- Show each factor as (10 + 3) and (10 + 2) in a rectangular array
- Show tens x tens
- Show tens x ones
- Show ones x tens
- Show ones x ones

Write out the place value format and add to get the final product.

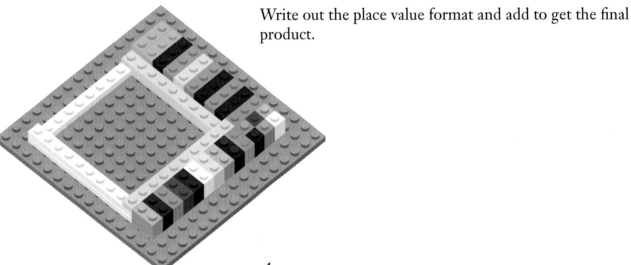

Answer:

10 x 10 = 100
10 x 3 =   30
2 x 10 =   20
2 x 3 =   6
_____

Product =156

2. Can you model the multiplication of 11 x 11?

Students should:
- Show each factor as (10 + 1) and (10 + 1) in a rectangular array
- Show tens x tens
- Show tens x ones
- Show ones x tens
- Show ones x ones

Write out the place value format and add to get the final product.

Answer:

10 x 10 = 100
10 x 1 =   10
1 x 10 =   10
1 x 1 =   1
_____

Product = 121

APPENDIX

Base Plate Paper

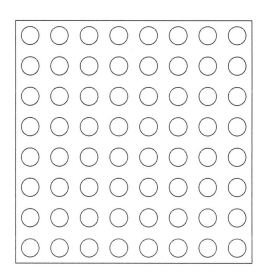

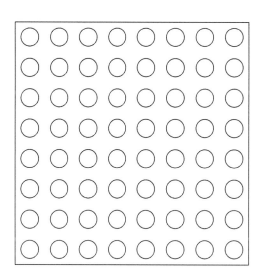

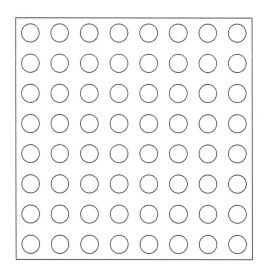

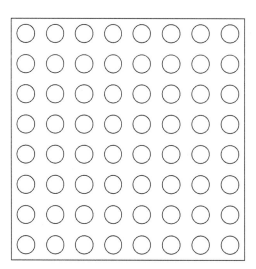

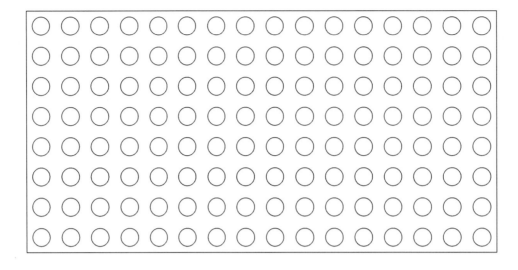

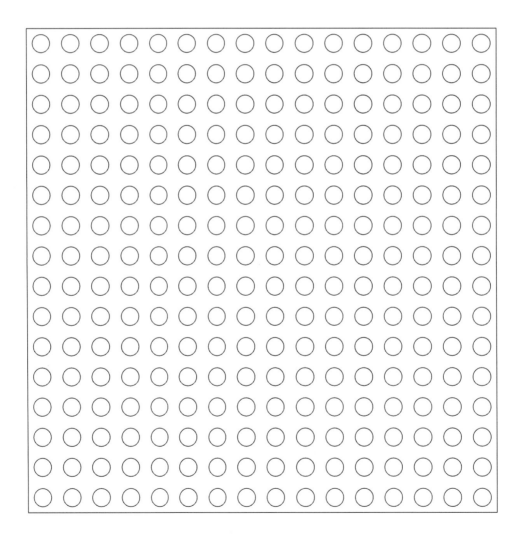

Made in the USA
Lexington, KY
01 August 2018